SWIMMING WITH SHARKS

FROM ADVENTURERS TO MARINE BIOLOGISTS

BY AMIE JANE LEAVITT

a Capstone company — publishers for children

Raintree is an imprint of Capstone Global Library Limited, a company incorporated in England and Wales having its registered office at 264 Banbury Road, Oxford, OX2 7DY – Registered company number: 6695582

www.raintree.co.uk
myorders@raintree.co.uk

Text © Capstone Global Library Limited 2023
Hardback edition published in 2023.
Paperback edition published in 2024.
The moral rights of the proprietor have been asserted.

All rights reserved. No part of this publication may be reproduced in any form or by any means (including photocopying or storing it in any medium by electronic means and whether or not transiently or incidentally to some other use of this publication) without the written permission of the copyright owner, except in accordance with the provisions of the Copyright, Designs and Patents Act 1988 or under the terms of a licence issued by the Copyright Licensing Agency, 5th Floor, Shackleton House, 4 Battle Bridge Lane, London SE1 2HX (www.cla.co.uk). Applications for the copyright owner's written permission should be addressed to the publisher.

Edited by Carrie Sheely
Designed by Dina Her
Original illustrations © Capstone Global Library Limited 2023
Picture research by Kelly Garvin
Production by Tori Abraham
Originated by Capstone Global Library Ltd
Printed and bound in China

978 1 3982 2287 8 (hardback)
978 1 3982 2286 1 (paperback)

British Library Cataloguing in Publication Data
A full catalogue record for this book is available from the British Library.

Acknowledgements
We would like to thank the following for permission to reproduce photographs: Alamy/Franco Banfi.Nature Picture Library, 16; BluePlanetArchive.com/Doug Perrine, 18; Getty Images: Joel Carillet, 26, VW Pics/Contributor, 27; Newscom: Franco Banfi/ZUMA Press, 11, HUGH GENTRY/REUTERS, 21, 28, Paul Souders Danita Delimont Photography, 13, Solent News/Splash News, 20; Science Source/Ted Kinsman, 7; Shutterstock: Andy Deitsch, 17 (top), Chainarong Phrammanee, 9, Cq photo juy, 1, frantisekhojdysz, cover, 15, Greg Amptman, 23, 24, 25, jimcatlinphotography.com, 17 (top middle), Karel Bartik, 5, Kristina Vackova, 8, Marin Voeller, 22, Matt9122, 17 (middle)(bottom middle), Max Topchii, 14, Rich Carey, 29, Simon Burt, 10, SofotoCool, 6, VisionDive, 12, wildestanimal, 17 (bottom)

Every effort has been made to contact copyright holders of material reproduced in this book. Any omissions will be rectified in subsequent printings if notice is given to the publisher.

All the internet addresses (URLs) given in this book were valid at the time of going to press. However, due to the dynamic nature of the internet, some addresses may have changed, or sites may have changed or ceased to exist since publication. While the author and publisher regret any inconvenience this may cause readers, no responsibility for any such changes can be accepted by either the author or the publisher.

CONTENTS

CHAPTER 1
An amazing sight 4

CHAPTER 2
Get to know sharks 6

CHAPTER 3
Shark tourism 12

CHAPTER 4
Studying sharks 18

CHAPTER 5
Capturing the shot 22

    Glossary 30
    Find out more 31
    Index 32

Words in **bold** are in the glossary.

CHAPTER 1

An amazing sight

Imagine you're on a scuba dive in the ocean. Aqua waves ripple overhead as you dive deeper and deeper into the warm, clear water.

Fish flit past as they glide along a **reef**. Schools of bright yellow butterfly fish, shimmering red parrotfish and striped sheepsheads come into view. You smile in delight.

But the real star of the show doesn't swim in a group. This is what you are here for. A lone tiger shark glides through the sparkling sea. Its long, silvery body looks as smooth as silk. Its dark grey dorsal fin sticks up from its back, slicing through the water like a knife. You know this is an experience you'll never forget.

A diver gets up close to a Caribbean reef shark.

In the water with sharks

For years, people have been swimming with sharks. They get up close to hammerheads, great whites, nurse sharks and more. Many people who swim with sharks study these majestic creatures. Others, such as tourists, do it just for fun!

CHAPTER 2

Get to know sharks

Sharks are fish that have been swimming in the seas since **prehistoric** times. Some of the first shark **fossils** date back 450 million years. They lived before dinosaurs.

Sharks have rows of teeth lined up behind one another.

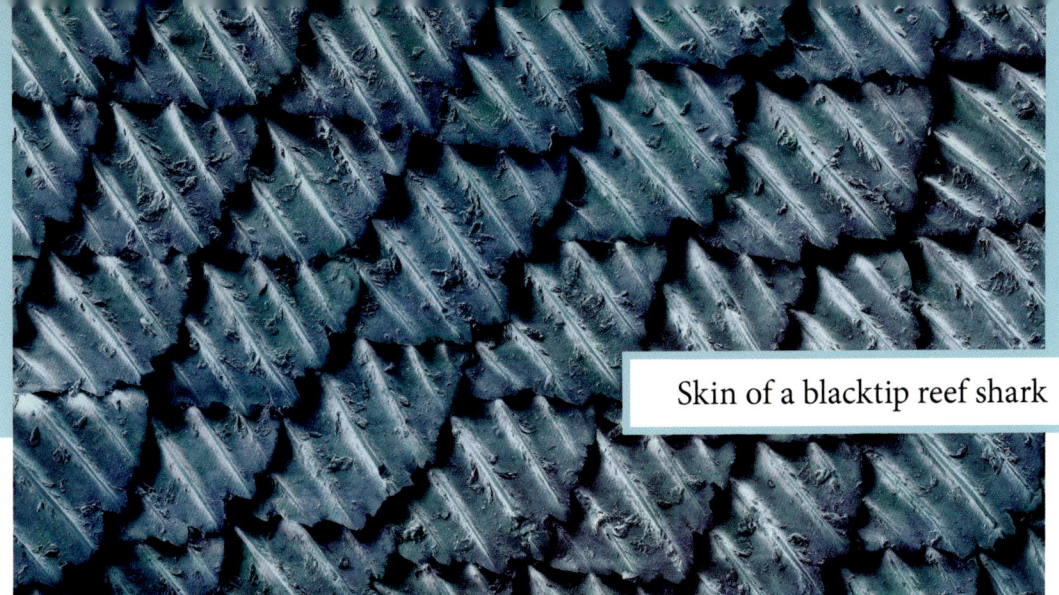

Skin of a blacktip reef shark

Shark bodies

Sharks have special skeletons that are different from many other fish. Most fish have skeletons made of bones. But shark skeletons are made of cartilage. This tissue is a bit softer and more flexible than bone.

Sharks have more body parts and abilities that set them apart from other fish. They have several rows of teeth. When one tooth breaks, a tooth from another row moves up to fill the empty space.

Shark skin is made up of tiny scales called **denticles**. These scales are flat and V-shaped. If you run your hand across shark skin, it feels smooth going in one direction, but like sandpaper going in the other direction.

A grey reef shark swims in the Pacific Ocean.

Sharks have good vision. Their eyes let in light to help them see at night or in the darkness at the bottom of the sea.

Sharks have two dorsal fins. One is near the centre of the back. It is shaped like a triangle. The other is smaller and is located near the tail.

Shark species

There are more than 500 **species** of sharks. The ocean is a big place, though, so there may be more shark types that haven't been discovered yet. The smallest known sharks are dwarf lantern sharks. Most of these sharks could fit in a person's hand. The biggest sharks are whale sharks. They can be up to 18 metres (60 feet) long.

FACT
Some sharks, such as makos and great whites, have to keep swimming to keep water running over their **gills** and breathe.

A whale shark swims through the sea. These sharks weigh about 20 tonnes.

Sharks mainly eat meat. Most hunt other animals for food, but a few species are filter feeders. These sharks suck in huge gulps of water. They filter the water through their gills. The sharks then eat tiny animals and plants called **plankton** that get trapped in their mouths.

A filter-feeding basking shark opens its mouth to take in water.

A Greenland shark swims under the ice near Canada.

Where sharks live

Sharks live in most parts of the ocean. They swim near the ocean floor and in shallow waters. They glide under ice and swim in the cold water near the Arctic. They do not swim near Antarctica.

Just as many birds fly south or north in different seasons, some sharks **migrate**. These sharks move from place to place at certain times of the year. Sharks migrate to find mates and to give birth. They also migrate in search of a better food supply.

CHAPTER 3

Shark tourism

Swimming with sharks is becoming more popular around the world. More than 575,000 people go on shark swimming or watching tours every year. On these tours, guides take swimmers to locations where certain types of sharks can be found.

Scuba divers watch great whites from inside a cage.

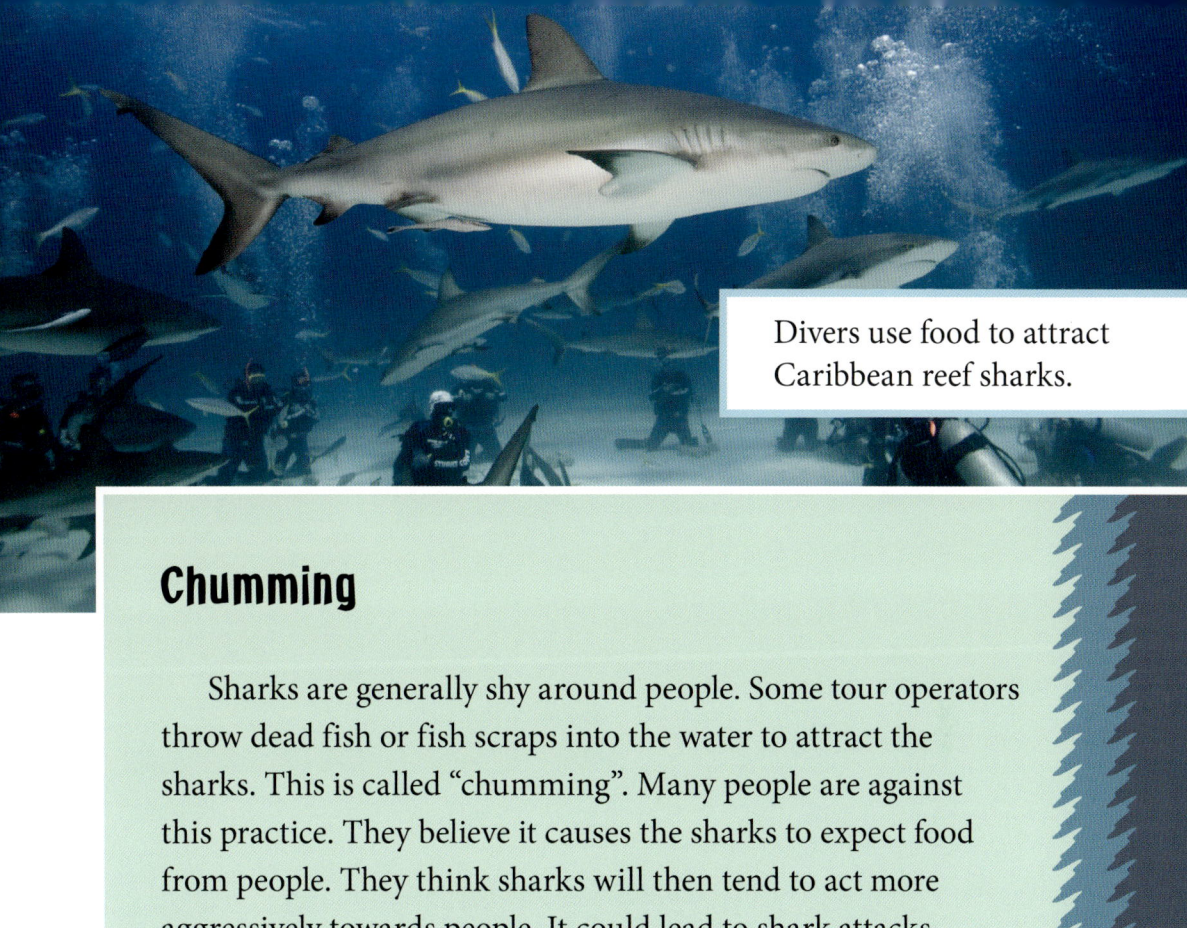

Divers use food to attract Caribbean reef sharks.

Chumming

Sharks are generally shy around people. Some tour operators throw dead fish or fish scraps into the water to attract the sharks. This is called "chumming". Many people are against this practice. They believe it causes the sharks to expect food from people. They think sharks will then tend to act more aggressively towards people. It could lead to shark attacks.

Cage diving

Some shark species are more aggressive and dangerous than others. The great white is one of them. Yet many people want to see this big-toothed animal up close. A cage is one solution. People go inside a big metal cage that is lowered into the water. From there, they can view the sharks in their natural environment while staying safer.

A girl snorkels near a whale shark.

Snorkelling and scuba diving tours

Snorkelling is another way people can swim with sharks on tours. Snorkels are tubes that stick up out of the water. They let swimmers breathe while their faces are in the water. People often snorkel near reefs and swim along with nurse sharks and reef sharks.

Snorkelling is also a common way to swim with whale sharks. Tour boats take groups to where these massive animals migrate. People often go to snorkel with whale sharks near Mexico's Yucatán Peninsula.

Scuba diving allows people to swim with sharks in deep water. One disadvantage to this is that it can be harder for swimmers to get close to sharks. Sharks can get scared when they see the tank and bubbles around a diver.

FACT

A study by a **conservation** group said that the shark diving industry in Florida, USA, supported more than 3,700 jobs in 2016.

Sharks swim above a group of scuba divers near the seabed.

People can scuba dive with sharks in many places around the world. The waters near Mexico, Hawaii, the Caribbean Islands, Australia and South Africa are popular places.

Free diving

People also **free dive** to see sharks. Free diving allows swimmers to go deep in the water without the use of air tanks. Free divers train to hold their breath for long periods of time. It takes a great deal of conditioning and practice to free dive.

A free diver (front) and a photographer get close to a blue shark.

SHARKS PEOPLE SWIM WITH

Type of shark	Size of shark	Top swimming locations
whale sharks	5.5 to 10 m (18 to 33 feet) long	Mexico, Galápagos Islands, Philippines, Maldives
nurse sharks	2.1 to 2.7 m (7 to 9 feet) long	Bahamas
Caribbean reef sharks	2 to 3 m (6.5 to 10 feet) long	Caribbean Islands, Bahamas
hammerhead sharks	0.6 to 6 m (2 to 20 feet) long	Bahamas, Galápagos Islands, Costa Rica, Mexico, French Polynesia
great white sharks	3.4 to 6 m (11 to 20 feet) long	Mexico, South Africa, Hawaii, Galápagos Islands, Australia, New Zealand

CHAPTER 4

Studying sharks

Many people who swim with sharks are professionals, including scientists, researchers and conservationists. They work to help expand what people know about these majestic swimmers. Conservationists help protect sharks and other marine life to ensure their survival.

A diver tags a Caribbean reef shark.

Tracking sharks

One way to learn more about sharks is to track them. Scientists attach digital tags onto a shark's dorsal fin or onto another body part. They often do this while swimming with the sharks. From a distance, they propel a tag out of a metal tube. The tag then attaches to the shark. The tag gathers data over a period of time and sends it back to the scientists.

What does this data tell the scientists? It allows them to track the paths of sharks. It shows them information about shark migration. This includes how far they travel in the ocean, where they feed and where they give birth.

FACT

Scientists tracked a whale shark in 2018 that travelled more than 20,000 kilometres (12,000 miles) to migrate. It was the longest migration journey ever recorded.

Ocean Ramsey observes sharks in the Bahamas.

Watching shark behaviour

Researchers learn about shark behaviour by swimming with these animals regularly. They learn how sharks interact with other sharks, how they hunt and how they take care of their young. Sharks live for 20 to 30 years. Migrating sharks often come back to the same places every year. This helps shark researchers who swim in those places get to know sharks as individuals.

Ocean Ramsey is a marine biologist in Hawaii. She studies shark behaviour in the wild. She has found that just like people, every shark is unique and different. Each one has its own personality.

One shark she has swum with often has been named Curly Girl. The shark was given this name because her dorsal fin has a tiny curl at the top. Curly Girl is a tiger shark.

Ocean Ramsey gives information to tourists before they go on a shark dive tour near Hawaii.

CHAPTER 5

Capturing the shot

Marine photographers and videographers are another group of professionals who swim with sharks. Their images and videos help researchers study sharks in the wild. They also help the public learn about sharks' important role in the ocean. Many of their images and videos are added to nature films, documentaries and science books.

Shark photographers and videographers often scuba dive. This allows them to stay in the water for long periods of time. It can take time to get equipment ready and to set up a shot.

A photographer gets a shot of a great hammerhead in the Bahamas.

A videographer records a lemon shark in the Bahamas.

Special equipment

A person can't take an ordinary camera into the ocean. It would get ruined in the water. Photographers use special waterproof cases for their equipment. These plastic coverings prevent water from getting inside and damaging the electronics.

Photographers often take devices called strobes into the water too. They produce light to help the photographer capture images in deeper ocean waters that get little sunlight.

A team photographs a tiger shark.

Teamwork

Shark photographers often work in teams. They work with other photographers and researchers. In places the teams are unfamiliar with, a local guide who knows the sea in that area might lead them.

As these teams work, they pay attention to sharks' behaviour. If a shark starts to act aggressively, they might need to back off and give the shark its space. By staying aware, they keep everyone on the dive safe.

To stay safe, photographers watch shark behaviour closely as they approach the large animals.

Helping to save sharks

Have you heard the phrase "A picture is worth a thousand words"? This is especially true when it comes to sharks.

Beautiful photos and videos of sharks help people change their ideas about sharks. Over the years, scary shark films have made many people afraid of sharks. But these films are fiction. Shark attacks on humans are extremely rare, and most shark attacks are not deadly.

If overfishing continues, some shark species may die out.

A diver inspects a dead hammerhead shark on the ocean floor that has been finned.

FACT

In 2019, there were only 64 unprovoked shark attacks worldwide. Of these, only two were fatal.

When people start to understand sharks, it's more likely they will want to protect them. Every year, humans kill about 100 million sharks. If **overfishing** continues, there won't be many sharks left. Some of these sharks are killed illegally in practices such as shark finning. This involves killing sharks for their fins. The fins are often used in shark fin soup.

Juan Oliphant (on boat) leads a shark tour in Hawaii.

Juan Oliphant is a shark photographer and conservationist. He wants to help change people's fear of sharks. He knows that the sharks cannot speak for themselves, so he hopes that his photographs speak for them. He has worked with conservation groups to help save sharks. He also helped start an organization that offers shark tours and supports shark conservation.

An amazing experience

Would you swim with sharks? What type of shark would you want to see the most? Would you see great hammerheads? Or would you rather come face-to-face with great whites? Many people say seeing these animals up close is the experience of a lifetime. Maybe it's one you will get to have one day!

A diver encounters a leopard shark.

Glossary

conservation protection of animals, plants and natural resources

denticle small, tooth-like scale that covers a shark's skin

fossil remains or traces of plants and animals that are preserved as rock

free dive swim underwater without scuba equipment and while holding your breath

gill body part on the side of a fish; fish use their gills to breathe

migrate travel from one area to another; some animals move from one area to another to find warmer weather

overfish fish too much, threatening the survival of a type of fish

plankton tiny plants and animals that drift in the sea

prehistoric from a time before history was recorded

reef strip of rock, coral or sand near the surface of the ocean

species group of animals or plants that share common characteristics

Find out more

Books

Fish (Animal Classification), Angela Royston (Raintree, 2015)

Shark: Killer King of the Ocean (Top of the Food Chain), Angela Royston (Raintree, 2019)

Sharks (DKfindout!), DK (DK Children, 2017)

Terrors from the Deep: True Stories of Surviving Shark Attacks (True Stories of Survival), Nel Yomtov (Raintree, 2015)

Websites

www.bbc.co.uk/cbbc/shows/shark-bites
Learn about many different types of sharks.

www.dkfindout.com/uk/animals-and-nature/fish/sharks
Find out more about sharks and take the Fish: true or false quiz!

Index

cage diving 13
cartilage 7
chumming 13
conservationists 18, 28

denticles 7
dorsal fins 4, 8, 19, 21
dwarf lantern sharks 9

filter feeders 10
finning 27
fossils 6
free diving 16

great white sharks 5, 9, 12, 13, 17, 29

hammerheads 5, 17, 22, 27, 29
Hawaii 16, 17, 21, 28

mako sharks 9
Mexico 14, 16, 17
migration 11, 14, 19, 20

nurse sharks 5, 14, 17

Oliphant, Juan 28
overfishing 26, 27

photographers 16, 22, 23, 25, 28
plankton 10

Ramsey, Ocean 20, 21
reef sharks 5, 8, 13, 14, 17, 18

sand sharks 14
scuba diving 4, 12, 15, 16, 22, 27
shark attacks (on people) 13, 26, 27
snorkelling 14
strobes 23

tags 18, 19
teeth 6, 7
tiger sharks 4, 21

videographers 22, 23
vision 8

whale sharks 9, 14, 17, 19